William N. Berkeley

An Investigation of the Relative Rate of Reduction of the

Isomeric Nitrobenzoic Acids

William N. Berkeley

An Investigation of the Relative Rate of Reduction of the Isomeric Nitrobenzoic Acids

ISBN/EAN: 9783337339883

Printed in Europe, USA, Canada, Australia, Japan

Cover: Foto ©berggeist007 / pixelio.de

More available books at **www.hansebooks.com**

AN INVESTIGATION

OF THE

RELATIVE RATE OF REDUCTION OF

NITROBENZOIC ACIDS.

A Dissertation presented to the Board
of University Studies of the Johns Hopkins
University for the Degree of Doctor of Phi-
losophy

by

William N. Berkeley

--o--

Baltimore, Maryland.
June, 1899.

Contents.

--

Acknowledgement.

This investigation was undertaken at the suggestion and prosecuted under the immediate supervision of Doctor Ira Remsen, to whose direction and counsel I gladly make this acknowledgement of my obligation.

To Doctor Morse my thanks are also due, for his guidance throughout the course in Analytical Chemistry, pursued under his supervision, and also for his advice on matters more directly connected with this investigation.

I am also indebted to Doctors Mathews and Ames for the benefits derived from their instruction and advice.

--ooo--

HISTORICAL.

----

(1)
In an article by H. Limpricht appearing in 1878, and
entitled "Reduction der Nitroverbindungen mit Zinnchlorür
und quantitative Bestimmung der Nitro Gruppe", he gives the
results of some investigations conducted by his two pupils,
Heinzelmann and Altmann, on the use of stannous chloride and
iodine solutions as a means for the quantitative determination of the nitro group in organic compounds.

Such determinations were made of the nitro group in
meta-nitro-sulphobenzoate of sodium and its ortho-isomer.
Also in para-nitro-sulphobenzoate of ammonium, ortho and
meta-nitro-benzoic acid, meta-nitrobenzoate of barium nitro-
brombenzoic acid, dinitro-sulphobenzoate of barium, nitro-
sulphobenzoate of potassium, meta-nitro-sulphobenzamide or-
tho-nitrophenol, meta-dinitrobenzene, nitro-bromsulphoben-
zoate of potassium and of ammonium, para-nitrotoluene, meta
nitroaniline, etc.

The results of these investigations were such as to
lead Limpricht to suggest the general applicability of the
method, not only for purely scientific purposes but also to
those of a technical nature. Later, in an article by S. H.
----
(1) Ber. XI., 35.

(1)

Young and R. C. Swain, "On the Volumetric determination of the Nitro group in organic compounds", the results of determinations of the nitro group in dinitrobenzene by means of stannous chloride and iodine solution are given, and the authors express the opinion that the method will prove generally applicable to other similar problems.

The article of Limpricht suggested the method as one that could possibly be used in the determination of the relative rate of reduction of the isomeric nitro bodies, and following this suggestion, an investigation was begun by Remsen and Burton at the Johns Hopkins University, On the relative rate of reduction of ortho, meta, and para-nitrobenzoic acids. No conclusive results were obtained and the work described in this paper is a continuation of that investigation, and like that, was undertaken at the suggestion and prosecuted under the supervision of Dr. Remsen, in the hope that the method mentioned might prove useful in the solution of this and similar problems.

MATERIAL USED AND ITS PREPARATION.

The substances studied, as already stated, were the three isomeric nitrobenzoic acids, it being originally intended to extend the investigation to a number of other ni-

----

(1) Jr. Amer. Chem. Soc., Oct. 1897.

tro bodies.

The ortho and meta acids were prepared according to
(1)
the method described by Levy with certain modifications
suggested by Mr. E. E. Reid, Fellow in Chemistry at this
University. The para acid, being obtained in a very small
percentage by this method, was prepared by another one,
(also recommended by Mr. Reid) which consisted in oxidizing para-nitro-toluene by means of sodium bichromate and
sulphuric acid. By means of repeated crystallizations
from water, to which a little potassium permanganate was
added, the three acids were obtained in a state of purity
as was shown by their constant and sharp melting-points.

METHOD OF PROCEEDURE.

Owing to the extreme sensitiveness of the stannous
chloride, to any change in the conditions under which the
determinations were made, no little difficulty was experi-
enced in maintaining from day to day such constancy in the
conditions as would furnish results reasonably concordant.

The stannous chloride used in the reduction was made
every day from the salt, crystallized just before each se-

----

(1) Anleitung zur Darstellung organisch-chemischer Papa--
rate, pp. 158, et.seq. (4th Edit.)

ries of determinations by quickly cooling a hot and concentrated solution by means of a freezing -mixture.

The iodine solution was kept in a metallic case to protect it as far as possible from all light, and was titrated against sodium hypo♀sulphite, which, in turn, was standardized every day or two with iodine very carefully dried, sublimed, and preserved in a desⓒiⓒator.

In the first determinations, the iodine_equivalent of five cubic centimetres of the freshly prepared and unheated stannous chloride solution was determined and from this number was subtracted the number of cubic centimetres of the iodine solution required by each sample of the acid after reduction, but it was soon seen that this method of proceedure made no allowance for the oxidation of the stannous chloride during the subsequent heating so this was changed, as will be described later. A great many sources of error were detected in the original method, and these were eliminated as far as possible.

The method after the elimination of a number of sources of error was as follows:-

Into three (3) flasks of about one hundred and twenty-five (125) cubic centimetres capacity, and each containing fifty (50) cubic centimeters of water, were weighed

from one to two-tenths (0.10 - 0.20) of a gramme of the isomeric acids. After the solution of the three (3) samples, these three (3) flasks, together with a fourth containing fifty (50) cubic centimeters of water alone, were transferred to a specially devised water-bath consisting of a smaller fitted into a larger one, the latter being heated by a Bunsen flame, which, when necessary, was controlled by a thermostat. [1]

After the flasks had been brought to the temperature at which it was proposed to determine the reduction, and the maintenance of this temperature was assured, fifty (50) cubic centimeters of an acid solution of stannous chloride, previously heated in a closed vessel to the same temperature, were added to each flask which was then securely closed with a stopper.

At intervals of one-half, one, one and a-half and two hours from the time at which the stannous chloride had been added to the respective flasks, samples containing approximately twenty-five (25) cubic centimeters were taken from each; quickly cooled by immersion in ice-water; ten cubic centimeters measured into a beaker, and after adding fifty (50) cubic centimeters of water, two (2) cubic centi-

----

(1) This thermostat was kindly made for me by Mr. C. E. Waters, lecture assistant.

(1)

meters of soda solution, and ten (10) drops of a dilute
starch solution, the excess of stannous chloride in each
sample was determined by means of a N 1/100 iodine solution.

From the number of cubic centimeters of the iodine
solution, representing the iodine equivalent of five (5)
cubic centimeters of the stannous chloride solution, heated
for the same time and under the same conditions of temperature and dilution as the sample of acid, was subtracted
the number of cubic centimeters of the iodine solution required by each sample after reduction.

The difference multiplied by the weight of the nitrobenzoic acid corresponding to each cubic centimetre of
the iodine solution gives the weight of the acid reduced
from which the percentage of reduction is readily calcula-
ted.

In the following tables are given the figures representing the percentage of reduction after the samples had
been heated at the temperature and for the time noted.

They represent the average of several hundred deter-
- -
(1) Na_2CO_3 180 gms.
      Rochelle Salt 240 gms.
      H_2O qs.ad. 1 Li

minations, those which were obviously inaccurate from any known irregularities in the determination being excluded.

Using a stannous chloride solution containing eight (8) grammes of the tin salt to the litre [1] -

RESULTS. HEATING THIRTY (30) MINUTES

Temperature	68° 70°	78°- 80°
Ortho acid	3.4%	12.1%
Meta "	1.0%	5.5%
Para "	10.0%	24.0%

HEATING SIXTY (60) MINUTES AT ABOVE TEMPERATURE

Ortho acid	12.1%	23.7%
Meta "	8.0%	17.4%
Para "	27.8%	39.0%

HEATING NINETY (90) MINUTES - SAME TEMPERATURE

Ortho acid	19.9%	33.6%
Meta "	13.0%	25.9%
Para "	34.1%	46.1%

----

[1] Two solutions were used at different times.
  (a) $SnCl_2$ 8 gms.   (b) $SnCl_2$ 16 gms.
      HCl   100c.c.   H Cl   100c.c.
      H_2O q.s.ad 1 Li. H_2O qs. ad 1 Li.

HEATING ONE HUNDRED AND TWENTY MINUTES

Temperature	68° - 70°	78° - 80°
Ortho acid	21.7%	38.5%
Meta "	16.9%	31.4%
Para "		52.2%

Numbering the intervals of thirty (30) minutes 1.2 and 3 and subtracting the percentages of reduction in successive equal intervals, we get the following differences

Temperature 68°- 78°

	I.	II.	III.
Ortho	8.7%	7.8%	1.8%
Meta	8.0%	5.0%	3.9%
Para	17.5%	6.3%	-

For a temperature of 78°- 80° we get:

	I.	II.	III.
Ortho	11.6%	9.9%	4.9%
Meta	11.9%	8.5%	5.9%
Para	15.0%	7.1%	6.1%

In a similar manner, determining the ratio of the percentages of the isomeric acids reduced in the equal in-

tervals of time, and at the same temperature, we get (omitting the figures for the first thirty (30) minutes) the ratio ortho:meta: para

Temperature	68^{0}- 70^{0}	78^{0}- 80^{0}
after sixty minutes	1.5: 1:3.5	1.4: 1: 2.2
" ninety "	1.5: 1:2.6	1.3: 1: 1.8
" one hundred and twenty minutes	1.3:1:-	1.2: 1: 1.7

Using a solution of stannous chloride twice as strong as the above, i.e., containing sixteen (16) gramms of the tin salt to the litre, we get the following

RESULTS - HEATING THIRTY (30) MINUTES

Temperature	60^{0}- 61^{0} -	70^{0}-71^{0}-	80^{0}- 81^{0} -	87^{0}- 38^{0}
Ortho acid	9.25%	20.4%	31.6%	42.6%
Meta "	2.8%	12.4%	-	38.3%
Para "	-	27.7%	55.6%	68.8%

HEATING ONE HOUR - TEMPERATURE AS ABOVE

Ortho acid	12.5%	34.6%	43.4%	56.1%
Meta "	3.9%	26.7%	37.6%	50.2%
Para "	-	40.7%	59.4%	79.0%

HEATING ONE AND ONE-HALF HOURS

Temperature	$60^0 - 61^0$ -	$70^0 - 71^0$ -	$80^0 - 81^0$ -	$87^0 - 88^0$
Ortho acid	21.2%	42.0%	50.0%	61.0%
Meta "	11.6%	32.5%	44.9%	56.4%
Para "	-	68.0%	-	80.4%

HEATING TWO (2) HOURS

Ortho acid	28.3%	-	54.8%	63.4%
Meta "	17.4%	-	50.2%	60.4%
Para "	-	-	69.8%	82.6%

As before numbering each successive interval of thirty
(30) minutes 1,2,3, and subtracting the figures representing the percentages of reduction in these successive intervals we get the following

RESULTS - TEMPERATURE   60 - 61

	I.	II.	III.
Ortho acid	3.25%	8.7%	7.1%
Meta "	11.0%	7.9%	5.8%

RESULTS - TEMPERATURE   70 - 71

	I.	II.	III.
Ortho acid	14.2%	7.4%	-
Meta "	14.3%	5.8%	-
Para "	13.0%	27.3%	-

RESULTS - TEMPERATURE 80 - 81

		I.	II.	III.
Ortho	acid	1.8%	6.6%	4.8%
Meta	"	-	7.3%	5.3%
Para	"	3.8%	-	-

RESULTS - TEMPERATURE 87 - 88

		I.	II.	III.
Ortho	acid	13.5%	4.9%	2.4%
Meta	"	11.9%	6.2%	4.0%
Para	"	10.2%	1.4%	2.6%

Comparing as before the figures representing the
percentage of reduction of each of the acids in each period
of thirty minutes we have the ratio ortho ; meta; para

TEMPERATURE - 60°- 61° - 70°- 71° - 80°- 81° - 87°- 38°

	60°-61°	70°-71°	80°-81°	87°-38°
Heating sixty minutes	3.2:1:-	1.3:1:1.5	1.1:1:1.6	1.1:1:1.6
" ninety "	1.8:1:-	1.3:1:1-	1:1:-	1+:1:1.4
" one hundred and twenty -	1.6:1:-	-	1-:1:1.4-	1+:1:1.4

The great irregularities noted in comparing the figures expressing the reduction in the early stages of the
reaction, are no doubt attributable to the fact that some
considerable time was necessary to get the reaction well

underway, and, as the figures show the meta acid to be less
susceptible to the action of the reducing action, and especially so at the lower temperature, this would naturally
present the greatest irregularity, a fact shown by the figures expressing the percentages of reduction.

By far the greater number of determinations were
made with the weaker solution of stannous chloride, and as
it was found that the reaction at temperatures below sixty
(60) degrees was very slow, (no reduction being detected
even at the end of two (2) hours in the case of the meta,
or at the end of one (1) hour in the case of the ortho acid)
subsequent determination were made at temperatures varying
from 70° - 38°

It will be seen from the description of the method
of procedure, that the figures representing the percentages
of reduction are necessarily based on the assumption that
the stannous chloride, in each of the flasks containing the
acids to be reduced, is <u>oxidized</u> to the same extent, in the
course of the heating, as it is in the flask containing the
stannous chloride alone; but while this assumption is necessarily made, it is not unreasonable to suppose that not
only the presence of the organic acid in the stannous chlo-
ide solution, but also the extent to which this is suscepti-

ble to reduction may influence the extent to which the stannous chloride is oxidized.

We might, therefore, conclude that the amount of reduction of the organic acid is determined to some extent by the greater or less loss in the efficiency of the tin salt through its oxidation, this in turn being influenced, as just stated, by the greater or less ease with which the acid in question undergoes reduction.

Some weight seems to attach to this hypothesis, from some observations made in regard to the rather peculiar behavior of the meta acid especially at lower temperatures.

It was found, for example, that in certain determinations samples of this acid, taken and titrated as usual after heating with the stannous chloride required more iodine than a sample of the tin salt heated alone.

The recurrence of this phenomenon throughout a whole series of samples, taken at intervals of one-half hour for two hours, precludes the possibility of assigning it to analytical errors.

This could be explained, ~~though~~ by the hypothesis that the tendency of the meta acid to be reduced had lessened the susceptibility of the tin salt to be oxidized

which would, therefore, undergo oxidation to a less extent than the stannous chloride heated alone.

The fact already stated, that the meta acid at the lower temperature seems very sluggish appears to indicate that the two opposing forces, tending to oxidation on the one hand and to reduction on the other, tend to neutralize one another until sufficient time has elapsed to start the reaction between the tin salt and the acid.

This apparent inertness of the meta acid is of course in complete accord with the general conclusions adduced from the whole series of determinations as to the less susceptibility of this acid to the reducing action.

These conclusions are based on the results of a long series of determinations performed with great care.

The figures given in this paper are decidedly surprising, viewed in the light of results obtained in investigations conducted along somewhat similar lines, from which it appeared that the presence of certain groups in a variety of organic compounds, which were in the ortho position relative to certain residues seemed to exert over these latter a protective influence to a greater or less extent, when sujected to the influence of certain classes of re-

agents.

REVIEW OF SIMILAR INVESTIGATIONS.

That the introduction of certain elementary or
compound molecules into organic bodies influenced their
stability has been long known but only such investigations
shall be discussed in this paper as are more immediately
related to the problems described therein.
(1)
As early as 1867 Vollrath called attention to the
marked stability of tri-chlor-xylene against the oxidizing
action of chromic acid.
(2)
In this same year, also, Beilstein and Kreusler
showed that when nitroxylene from coal-tar xylene was oxidized it formed nitrotoluic acid, and they add, in no case
is the second methyl group of nitroxylene attacked by this
oxidizing agent.

(3)
In 1871, Rudolph Fittig in an article entitled
                                          ʌ
"Gesetzmässigkeiten in der Aromatischen Gru/ppe" states
that it seems at least very probable that all ortho compounds,when treated with chromic acid mixture, do not furnish the oxidation-products corresponding to the analogous
meta and para bodies but are entirely decomposed ,and he

adds that Beilstein had found that the oxygen, chlorine,
iodine and nitro derivatives of toluene treated with chromic acid mixture do not form the corresponding benzoic
acids but are burnt up.

(1)

Again, in 1878, E. v.Gerichten showed that when
chlor-cymene made from carvacrol [structure] is oxidized it
passes to chlor-toluic acid in which the position of
the chlorine is ortho to the methyl and meta to the propyl
group, the latter only being oxidized to the carboxyl group.

(2)

Schmitz, also, in 1878, in studying nitro-mesity-
lenic acid found that the two methyl groups which remained
unoxidized were in the ortho position relative to the nitro group and in brom-mfsitylenic acid the same relative
positions was proved.

In the mean time there appeared in 1873 the first
(3)
of a long series of articles by Dr. Ira Remsen, then at
Williams College, in which are given the results of a long
series of investigations, conducted for the most part by
Dr. Remsen and his students at the Johns Hopkins University,
and in which the whole question of the protective influence
of certain negative groups in an ortho position relative to

----

(1) Ber.XI., 364.   (3) Am.Jour.Sci., V. 354.
(2) A. 193,160.

certain oxidizable residues, was very elaborately investi-
gated, the results being published in a series of articles
under the general heading "On the Oxidation of Substitution
Products of Aromatic Hydrocarbons".

The conclusions drawn from these investigations
(1)
are expressed thus:

"The results thus far reached all agree, and they
make the conclusion extremely probable, that in all cases
now on record, in which hydrocarbon residues are shown to
be protected from oxidation by the presence of negative
groups, the latter are in the ortho position with reference
to the former, whereas, oxidizable residues when situated
in the meta or para positions with reference to the negative groups are, under the same circumstances, transformed
just as if the negative groups were not present."

Some of the facts upon which these conclusions
were based are these:
First, it was found that when a mixture of ortho and para-
totuene sulphonic acids was subjected to the oxidizing ac-
(2)
tion of chromic acid mixture, it yielded no orthosulphoben-
zoic acid, and when the potassium salts of ortho and para-

toluenesulphonic acids were treated separately in the same
way, the action in the case of the para compound was prompt
and vigorous, while in the case of the ortho isomer it was
sluggish and weak.

Again, on oxidizing orthotoluenesulphamide mixe-
ture with a little of the para isomer [1] the ortho was not at
first affected, while the second passed into the corresponding sulphaminebenzoic acid, and continued action only
served (to partly destroy the ortho body.

Later it was shown that by the oxidation of meta-
xylenesulphamide [2] an acid was formed whose formula was
shown to be HO_3C ⟨ $C-CH_3$ ⟩ in which the methyl group ortho
to the sulph- CO_2H oxic acid residue is protected
from oxidation.

The correctness of the above formula was shown,
first; by its conversion into oxy-toluic acid [3] whose formula is $HO \cdot C$ ⟨ $C-CH_3$ ⟩ as is proved by its conversion into
ortho cresol [4] ⟨ $C.CO_2H$ ⟩ and further the sulphamine metatoluic
acid was transformed into xylidinic acid [5] HO_2C ⟨ $C-CH_3$ ⟩ CO_2H

---

(1)   Am. Chem. Jour. I., 32 et.seq.
(2)   Am. Chem. Jour.I. 37.
(3)   Ibid. I., 48.
(4)   Ibid. I., 114.
(5)   Ibid. I., 119.

and also to isophthallic acid.

(1)

In a later article by Remsen and Morse they show
that when brom-ethyl-toluene is oxidized by chromic acid
mixture, it passes first to brom-ethyl-para-toluic acid,
and then to para toluic acid ,and the authors think that the
resistance offered by the methyl group to the oxidizing
agent is due to the fact that it is in an ortho position
relative to the bromine.

(2)

Afterwards it was shown by Remsen and Hall that
para-xylenesulphamide is oxidized to sulphamine-para-toluic
acid, and while the author did not prove directly the relative positions of the group in this compound, they conclu-
ded that the unoxidized methyl group was in an ortho position relative to the sulphamine group from its analogy to
cymene-sulphanide which is oxidized first to sulphamine-
para-toluic acid ,and then to the acid described by E. v.
Gerichten under the names of a-oxy-toluic-acid in which, as
he showed, the hydroxyl group is in the ortho position relative to the methyl.

On the other hand, Remsen and Hall showed that
mesitylene-sulphamide is oxidized to a sulphinide, and rea-

----

(1)   Am. Chem. Jour. I., 138.
(2)   Am. Chem. Jour. II., 50.

soning by analogy with other sulphinides, the acid from which the sulphinide came directly by elimination of water, must have had the carboxyl and sulphamide groups in the ortho position relative to one another. This, therefore, seems opposed to the theory of protection.

It was found later that cymene_sulphonic acid is converted by chromic acid mixture to an acid thought to be
(1)
identical with an acid described by Flesch which in turn passes into the oxy-toluic-acid of v. Gerichten.

(2)
Remsen and Kuhara describe the preparation of sulpho para toluic acid from para brom toluene, the transformations being represented graphically in this way

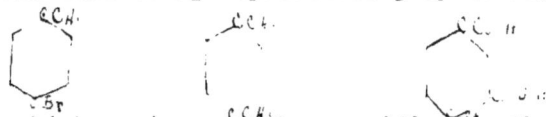

in which as is seen the unoxidized methyl group is ortho to the sulphonic acid residue.

Studying the influence of nitro groups ortho to
(3)
the oxidizable residue, they showed that nitrotoluic acid, when first converted into the corresponding amido-toluic acid, passes through the diazo compound into oxy toluic acid, which they showed to be identical with the oxy-toluic

----

(1) Ber. VI. 481.
(2) Am. Chem. Jour., 413.   (3) Am. Chem. Jour. III. 424

acid prepared by Remsen and Hall, and with the ortho-homo-

para oxybenzoic acid of Tiemann and Schotten $C_6H_3\begin{cases} OH & (1) \\ CH_3 & (2) \\ CO_2H & (4) \end{cases}$,

from which it followed that the nitrotoluic acid is

$C_6H_3\begin{cases} NO_2 & (1) \\ CH_3 & (2) \\ CO_2H \end{cases}$, in which the oxidized and unoxidized methyl

group is para and ortho respectively to the nitro group.

In commenting on this last investigation on the influence of nitro groups on an oxidizable residue the authors say:

"The facts thus far determined are not sufficient to
warrant a satisfactory conclusion though they make an af-
firmative answer (i.e., one affirming the law of protection) probable, at least for such substances as contain two
oxidizable residues together with the nitro group."

Finally, in an investigation conducted by Remsen
and Noyes [2] with the object of discovering whether the protective influence of negative groups extended to groups
more complex than methyl, they find that starting with di-
ethyl benzene sulphamide they get, after a series of transformations, finally sulphamine ethyl benzoic acid in which
the sulphamine group was thought to be in the meta-position
relative to the carboxyl.

Remsen and Day [3] also studied the effect of neg-

----

(1) B. XI., 767.   (2) Am. Chem. Jour. III., 20.
(3) Am. Chem. Jour. V. 148; IV. 197.

ative groups on propyl groups and found that by oxidizing

oxy-brom-cymene-sulphamide C_6H_1 $(C_3H_7$ $(SO, NH_2 (O)$ it passes to
$(C_3H_.$
$(CH_3 (P)$

C_6H_3 $(SO, NH_2 (O)$
$(CO_2H(P)$ (1)

On the other hand, Remsen and Keiser found that by the oxidation of para-dipropyl-benzoic sulphamide the protection of the propyl group is not complete.

Among a great many others who worked on problems similar to the one just described may be mentioned Richard
(2)
Meyer and A. Bauer who found that potassium permanganate converted cymene sulphonic acid [structure: $C CH_3$, $C SO_3H$] into [structure: $C CO_2H$, $C SO_3H$] while with the use of nitric acid it formed [structure: $C CH_3$, $C SO_3H$, $C C_3H_7$] which Meyer cites as a confirmation of Rem- [structure: $C CO_2H$ (3)] sen's protection theory

Prof. K. Shimomura of Kyoto, Japan, continuing some work commenced at the Johns Hopkins University found that para toluene-sulph-anilide [structure: $C CH_3$, $C CO_2H$, NHC_6H_5] when oxidized passes to [structure: SO_2NH_2] while the ortho-isomer gives [structure: $C CH_3$, $C SO_2$] as the the chief product, these results being in accord with results obtained by oxidizing para- and ortho-toluene-sulphamide.

----

(1) Am. Chem. Jour. V., 160. (2) A. 220,6.
(3) Private communication to Prof. Remsen.

23

(1)

W. A. Noyes also in an article entitled "On oxidation of benzene derivatives by potassium ferricymide," says
that all the work done on the oxidation of aromatic hydrocarbon derivatives leads to these conclusions: (1) chromic
acid does not oxidize ortho groups or only to a slight extent. (2) Nitric acid can oxidize ortho groups but prefers
meta or para groups.   (3) Caustic potash, in the case of the
only class of substances studied, oxidizes the groups ortho
to the negative group.   (4) Potassium permanganate in alkaline solution oxidizes ortho, meta and para groups pre-
fering the para in some cases.

Noyes, as a result of his own experiments with potassium ferricyanide in alkaline solution, found that by
treating ortho nitro toluene mixed with a little of the
para isomer he got ortho. and para-nitro benzoic acid.

He finds that para brom toluene is oxidized with
great difficulty and the ortho body with still more, if at
all. He also notes that more than two and a half times as
much para nitro toluene is oxidized than of toluene itself.

Another very important investigation, similar to
those described above, is that of Victor Meyer and his co-
workers the results of which were published in a series of
----
(1) Am. Chem. Jour. V., 97. (VII., 145; VIII., 167,176;
    X., 472)

articles appearing in the Berichte of the German Chemical
Society under the title of "Das Gesetz der Esterbildung der
Aromatischen säuren". (1)

The results obtained tend to support the hypothesis of the protective influence of the ortho position.

Meyer was lead to undertake this investigation by
his discovery of the fact that mesitylene carbonic acid when
heated with methyl alcohol and hydrochloric acid gave no
ester.

From his study of a large number of acids, he was
lead to the conclusion that whenever the carboxyl group of
the acid is in the ortho position relative to two other
groups, e.g., NO_2, CH_3, , OH CO_2H etc., they lose entirely or in part their capacity for forming esters in the
usual manner, i.e., by treating it with an alcohol and a
mineral acid.

It will be noted that this complete protection of
the carboxyl group occurs only when it is ortho to two other groups.

The problem involved in Meyer's work, as in that
of the other investigators mentioned, differs very essen-
tially in the nature of the reaction studied in these cases
from that one studied and described in this paper.

(1) B. 27,510,1580,3146-28,182,1254,2773,3197-29,839.

25

It also differs in this respect that, while in the former cases it was generally a negative group that seemed to exert the protective influence, in the latter, it is the negative group that is attacked by the reagent.

Some of the facts upon which Meyer based his conclusions were as follows:

First, as already stated, he found that mesilylene-carbonic acid in which the carboxyl is ortho to two methyl groups gave no ester when treated in the usual way while Duryl acid its unsymetrical isomer yielded ninety per cent.

Again, trinitrobenzoic acid   gave no ester.

Studying the effect of the hydroxyl group on ester forming capacity of an acid, he found that salicylic acid, an ortho compound, gave ninety per cent of an ester only by heating (about 70°) while the para isomer yielded an ester with ease and the meta more slowly. On the other hand, thymotinic acid   is with difficulty made to yield an ester.

They conclude that in the case of the hydroxyl group, the protective theory can be applied only in a limited way.

26

Some of the trihydroxy acids form esters while phloroglucinic acid liberates carbon dioxide and forms an ether of phloroglucin.

Meyer tried to discover the cause of the non-esterification of the acids of the general formula $R\langle\begin{smallmatrix}CO_2H\\CH\end{smallmatrix}$ ($R = CH_3.NO_2$, Br, OH, CO_2H etc.) which he thought due to one of two causes; either to the presence of some or all of the substituents, or to the unreplaced hydrogen.

To decide these points they studied the acids formed by further substitution of the trisubstituted acids.

It was found that tetra-brombenzoic acid$_{Br}\langle\begin{smallmatrix}CO_2H\\C.Br\end{smallmatrix}$ gave no ester nor did the acid$_{Br}\langle\begin{smallmatrix}CO_2H\\C.Br\end{smallmatrix}$ nor $\langle\begin{smallmatrix}\\C.Br\\C.Br\end{smallmatrix}$

The dibrom-benzoic acid $_{C.Br}\langle\begin{smallmatrix}CNH_4\\\end{smallmatrix}$ $\langle\begin{smallmatrix}CO_2H\\C.Br\end{smallmatrix}$ gave ninety per cent of an ester while its isomer $_{Br}\langle\begin{smallmatrix}CO.H\\C.Br\end{smallmatrix}$ gave none. $\qquad C.Br$

The conclusions drawn from these investigations
(1)
are expressed in these words:

"According to this, all questions that we have proposed above are experimentally answered. The results are unmistakable. As soon as two hydrogen atoms adjacent to a carboxyl are replaced by radicals, e.g., CH_3, Br, NO_2, etc, there results an acid which is not capable of forming an

----

(1) B. 27, 1585.

ester. The occurrence of the three substituents as well as
of the two hydrogen atoms in trisubstituted benzoic acids
is without influence on the phenomena".

Meyer proposes a rather unique hypothesis to explain why those acids which do not form esters in the usual
way, that is, with alcohol and a mineral acid, do so when
the silver salt of the acid is treated with an alkyl ha-
lide.

He thinks that when the silver atom enters the
carboxyl group it, by virtue of its superior size, forces
the adjacent groups aside to such a distance that they can
not exert their protective influence.

As evidence confirming this view, he offers the
following facts:-

He found that mesitylene acetic acid i.e.,
where the carboxyl group is removed from the immediate neighborhood of the methyl groups, forms ninety-six
(96) per cent of an ester while, as already stated, mesitylene carbonic acid   forms none.

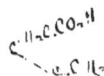

Mesitylene glyoxylic acid
also forms ninety-six percent (96%)
of an ester.

28

It was also found that the carboxyl group itself
exerts this protective influence where two are in an ortho
position relative to a third as in mellitic acid
which gives no ester. (Baeyer, on the other hand, found
that hexahydro mellitic acid is capable of esterrification)
Pegro-mellitic acid also gives ninety-six percentage of an ester. Premellitic acid
forms only a di-ester.

Investigating the two isomeric nitro phthallic acids
and he found that the former
gave a mono ester while the latter gave none.

(2)

Beilstein and Kurbatang had shown he says that
the acid forms a mono ester, while, on the other hand, they confirmed Graeben's statement that tetra
chlorphthallic acid gave an acid ester, while, according to
the theory, it should not form any. That this was not due
to the presence of the chlorine was shown by the facts that
the chlorbenzoic acid gave no ester while the
isomer did.

occur in two forms yielding two esters of the general for-

mula C_6Cl_4
```
       -COOR                OK    OK
                and
        COOR             C6Cl4
```

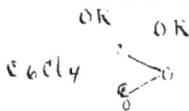

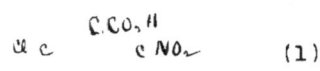

          (1)

      Meyer also studied the acid

to see if an acid in which the carboxyl group is ortho to

two <u>unlike</u> negative groups is capable of esterfication. He

found that in this case no ester is formed. He also found

that an acid of the general formula 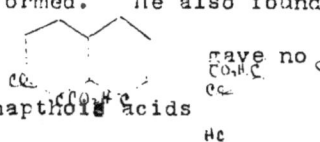 gave no

ester. Meyer studied the two chlor naptholi acids

         and found that the first gave no ester

while the second one gave ninety per cent (90%) of an ester

                   (2)

      Studying the rate of esterfication of isomeric

bodies he found the reaction was slowest in the case of the

ortho compounds and the same thing was proved in regard to

the rate of saponification.

               (3)

      In a later article though published with the

above Meyer said that Behla had shown that the acid

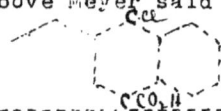 gave no ester as was to be expected while

----

(1) B. 28, 182.
(2) B. 28, 188.
(3) B. 28
(4) Carbon al   situated thus seem to act as substit-
    uents.
(5) B. 20, 703.

30

Meyer himself proved that the chlorine-free acid also gave none.

Meyer, in a subsequent article, modifies his former views as to the influence of hydroxyl groups on esterification, and says that, if two positions ortho to a carboxyl are possessed by substituents, and if one of these is an hydroxyl, no ester is formed in the cold.

As examples of these he cites phenyl salicylic acid C $C.CO_2H$ $C.OH$ and the acid represented by this formula neither of which gives ester while the isomer of the latter one does, i.e., $C.CO_2H$ $C.CO_2H$

Meyer also gives as further proofs of his stereochemical hypothesis which he had offered to explain why acids in which the carboxyl group was immediately adjacent to two negative groups form no ester while they do if the carboxyl is separated from these by one or atoms or atom-groups the following facts:-

Representing the group $C_6H_2R_3CO_2H$ (in which R = Radical) by ms, i.e. mesitylene-like, he found that Ms C .CO_2H. Ms CO C OH CO.H and Ms $CH_2CH_2CO_2H$ all gave esters.

Meyer also studied the difference in the effect

----

(1) B 20, 703.

of various substituents on ester-forming especially with
reference to the relative weights of the substituent as to
as to the relative intensity of their negative or other
character. With reference to the first he says: "those
radicals which prevented the formation of ester even under
the influence of heat were found to have a far greater
weight than those which prevented it only in the cold" As
examples of the first class he cites the groups Cl, NO_2, Br,
while in the second class he puts CH_3 and OH.

He also found that the esters of the acids
and $C.CO_2H$ behave very differently towards sapon-
ifying agents and states that "this surprising result shows
distinctly that those esters that are formed with difficulty are saponified with still greater difficulty and <u>vice
versa</u>".

Studying the rate of esterfication he found that
in the cases of ortho-, meta-, and paratoluic acids, ortho-,
meta-, and parabrom-benzoic acids and ortho- and metanitro-
benzoic acids, the ortho isomers were acted on much less
readily while experiments on the rate of saponification of
ortho- and metabrombenzoic acid esters gave similar results.
- -
(•) He proved that carbon atoms so situated act as negative
groups as regards their effect on ester-forming.

32

(1)

In an article by J. J. Sudborough on "Di-ortho
substituted benzoic acids" he describes the results obtained in an investigation, first, on the relative stability of
isomeric substituted benzoyl chlorides towards water and
caustic soda and secondly, on the behavior of isomeric aromatic nitrites and acid amides towards hydrolyzing agents.

Under the first head he gives the results of his study of
ortho-,meta -,and parabrombenzoyl chloride of 2-4, 3-5, 2-6
dibrombenzoyl chloride of 2-4-6, and 3-4-5 tribrombenzoyl
chloride and of 2-3-4-6 tetrabrombenzoyl chloride.

His conclusions based on the results of this investigation are stated thus: "The ortho chlorides in which
the substitution does not occur in either of the ortho positions are readily decomposed by dilute alkalies, the de-
composition being almost instantaneous at the boiling-point
of the alkali solution.

(2) The acid chlorides which have a bromine atom in one ortho position are relatively more stable towards alkalies
• • • • • •

(3) The acid chlorides in which both ortho positions are
occupied by bromine atoms are only converted into the corresponding sodium salt of the acid by long continued boil-
----
(1)   Chem. Soc. Jour. (Trans) 67,587,601.

ing with an alkali solution.

(4) If substitution has already taken place in the two ortho positions with respect to the $COCl_2$ group, the introduction of bromine atoms into the benzene nucleus seems to render the acid chloride still more stable."

In the second part of his article, Sudborough gives the results of a study of the behavior of 2-6 and 2-4 dibrombenzamide towards 80% sulphuric acid and says: the 2-4 amide differs from the 2-6 isomer in the readiness with which it is converted into the corresponding acid, the latter being characterized by its great stability towards hydrolyzing agents.

Claus and his pupils working on the hydrolysis of various aromatic nitrites [1] give results which for the most part accord with those of Sudborough.

Their experiments show that while most nitrites are hydrolyzed to the corresponding acids by boiling for several hours with 60%-70% sulphuric acid, those in which substitution occurs in the two ortho positions yield the acid amide and not the acid under the same conditions. Two important exceptions though apparently are 2-6 dichlor-benzonitrite and 2-4-6 methyl-chloronitrobenzo-nitrite".

----

(1) Jnl. fir Chemie. (2) 37,197, A. 265,266,269,274.

34

Wegschider also did some work on esterfication chiefly on hemipinic and opianic acids and decides that the difference in the behavior of acids towards alcohol and a mineral acid and that of their salt (Ag) towards alkyl halides is due to the fact that in the first case there is an addition of the acid and that this addition is more interfered with by the neighboring ortho groups than is the substitution of the alkyl for the metal.

This theory of the addition of the mineral acid[1] though, seems to have been advanced first by Henry,[1] Friedel[2] thought that in making an ester by an alcohol and hydrochloric acid, a chloride was first formed and that the necessary splitting off of the hydroxyl of the organic acid was prevented by the presence of the ortho groups.

One of the most important and fruitful investigations dealing with this subject of the protective influence of ortho groups is one that was conducted by Remsen and Reid at the Johns Hopkins University "On the hydrolysis of the acid amides and their derivatives" the results of which have just been published.

The results of this investigation which was conducted with the greatest skill and was characterized by an

---

(1) B. 10, 2041   (2) Zeitsch für Chemie. 12,488 (1869)

attention to the most minute details rarely found in similar work fully justified all the labor involved and furnish-
ed a great deal of evidence in support of the "protective
theory".

Among the substances studied were: benzamide and
its amido, chlor, hydroxy, brom methoxy, and ethoxy derivatives and the isomeric toluic amides.

The results of hydrolysis both with acids and alkalies went to show the very much greater stability of the
ortho isomer towards these reagents.

About the same time that the above investigation
by Remsen and Reid was being prosecuted, Kellas published
     (1)
an article   in which he gives the results of an investigation on the rate of esterfication of certain organic acids.

He endeavored to discover (1) some definite rule
for the rate of esterfication of the ortho, meta and para
acids - (2) Influence of temperature. (3) Influence of the
atom or atom group on the rate.

He studied: ortho, meta, and para-toluic acids, ortho, meta, and para-, nitro-chlor, brom, and iodo-benzoic acids, ortho, meta and para-oxy-benzoic acids.

---- ---
(1) Zeitschf für Physikalische Chemie. XXIV., 221.

36

Studying first chlor-brom-iodo-and nitro-benzoic
acids he found that the rapidity of the rate of esterification was always lost in case of the ortho isomer while the
relation between the meta and para compounds changed more
or less with the substituents. The isomeric toluic acids
gave similar results. This relation was also maintained
fairly well with changes in temperature.

Comparing the relative rates of reduction of all
chlor, brom, nitro-,and iodo-derivatives he found that the
nitro had by <u>far</u> the greatest retarding influence while in
the case of the iodo, brom, and chlor-derivatives the retar-
dation diminished in the order named.

Experiments on rate of saponification gave similar
results.

GENERAL CONCLUSIONS DRAWN.

It is thus seen that by far the greater number of
the experiments cited tend to show that the presence of
certain groups in an organic compound occupying an ortho
position relative to certain other groups affect in a greater or lesser degree their susceptibility to the action of
certain classes of reagents.

That this protective influence varies in degree
and kind with the nature of the reagent employed seems to
be proved by a number of the investigations noted as well
as by that forming the subject of this paper. While the
results of this last investigation were not as satisfactory
as it was hoped would be obtained, the constancy of the in-
dications leaves no doubt as to the correctness of the conclusions reached, i.e., that the three isomeric nitrobenzoic acids are, under the influence of an acid solution of
stannous chloride reduced at a rate increasing from the
meta to the para acid. The fact that with an increasing
time and temperature the ratio of the amount of the ortho
acid reduced to that of the meta acid diminishes is no
doubt attributable to the more rapid loss of the reducing
efficiency of the stannous chloride in contact with the
more readily reduced organic acid.

Biography.

-o-

The author of this dissertation was born at Aldie, Loudon county, Virginia, September 24, 1867.

His earlier education was obtained by attendance at the public schools of Albemarle county, Virginia, and later at public (High) Schools of Staunton, Virginia and at a private school at the same place.

He attended the University of Virginia, 1887-1888, 1891-2, and also for three summer sessions. He graduated there in the schools of Chemistry, Geology and Mineralogy, and for the year subsequent to his graduation was employed as Analytical Chemist at an Iron furnace in Pulaski City, Virginia.

In 1893 he was elected Professor of Chemistry at St. John's College, Annapolis, Maryland and held that chair till 1896, when the degree of B.S. (Hon.) was conferred on him.

In 1896 he entered the Johns Hopkins University as a graduate student, and has for the past three sessions pursued courses in Chemistry, Geology and Physics, holding an ordinary Virginia Scholarship throughout this period.